HIS BECKONING CALL:
YOUR HEART IS GOD'S DESIRE

By
Susan Stover

COPYRIGHT

Published by Hemingway publishers

Cover design by Hemingway publishers

ISBN: Printed in United States

HEMINGWAY PUBLISHERS

Table of Contents

INTRODUCTION

All one needs to do these days is look around to see what a terrible mess humans have made of this world. In my time on earth, I have witnessed such a decline in what we commonly call "human decency." It seems that there just are no longer any lines over which we are afraid, or too courteous, to cross when it comes to insisting on *our way, our opinion, and our rights*! Social media certainly fosters this mindset because of the inherent anonymity of this platform; one can hide behind their profile (which may or may not even represent the true person) and not worry about any repercussions. This fact really shows how fractured our societal expectations have become.

Look at our shopping centers, our movie theatres, or anywhere people are together in a large space. We must be mindful that there could be a gunman, an explosion, or even an out-of-control vehicle to take away from just having a relaxing, fun outing. Our guard must always be up. And we must be concerned about the safety and well-being of our children whenever they are out of our sight. There seems to be no "safe place" anymore.

Looking at this on an even larger scale, there is so much infighting among our elected officials and so much vitriol! There is no common ground between people of differing opinions. Everything is painted black or white; no room for tolerance. The Middle East and Ukraine have such hatred poured out upon them; ordinary citizens, the elderly, and children are put in the crosshairs without a second thought.

But lest we tend to look around us in fear and anxiety, I want you to know that we have a God Who loves us, is ever-present with us, and is not only able but is willing to help us! *God is beckoning to us!* He is saying, Come away, child! I will ease, relieve, and refresh your soul! (Matthew 11:28) Bring me your burdens and cares. "Look to Me and be saved, all the ends of the earth! For I am God, and there is no other." (Isaiah 45:22)

Beckon: to call or summon by a silent gesture

PREFACE

Ever since God created the universe and populated the earth with animals, vegetation, and humans, He has been active, present, and involved with His creation. There has never been a moment during which His attention was off us. In this book I will **show you how to recognize His presence**, and **how to hear *His beckoning call*.** I assure you that God sees you and is deeply in love with you! I pray that you will come to understand God's heart for you, and to believe that *your heart is His desire.* This is true no matter who you are, what you have done, or where you are currently in life. His love is constant, and His intention of reconciliation is as strong as it has ever been!

God is love; I'm sure you have probably heard this phrase at least once. Like anything that we hear often, or without context, it can easily become trite and meaningless. Truly knowing what these three words mean requires knowing about God, and knowing what true love really is. Neither of these comes easily; certainly, it takes life experiences to have a backdrop for true comprehension. My sincerest prayer is for the reader of this book to develop a deep, personal relationship with God, knowing the freedom that comes from being truly and deeply loved!

"The Lord appeared...saying, Yes, I have loved you with an everlasting love; therefore, with loving- kindness have I drawn you and continued My faithfulness to you."

(Jeremiah 31:3)

"For I will (fully) satisfy the weary soul, and I will replenish every languishing and sorrowful person."

(Jeremiah 31:25)

DEDICATION

I dedicate this book to all my spiritual mentors, Christian friends, and my family. You have all taught me something along the way. It is through ours and others' brokenness that we recognize our need for a Savior. It is through the fellowship of other believers that we learn of and begin to receive grace and mercy for our weaknesses and failings. Unless we see and experience the pain, chaos, and rejection of this world, we cannot truly appreciate what God has to offer us. It is in reacting to the lostness of our own souls that we will search for "something more." My prayer is that the reader will understand that *what we search for, and ultimately surrender to, matters!*

"For God so greatly loved and dearly prized the world that He (even) gave up His only begotten (unique) Son, so that whoever believes in (trusts in, clings to, relies on) Him shall not perish (come to destruction, be lost) but have eternal (everlasting) life."

(John 3:16)

"The Spirit of the Lord (is) upon Me, because He has anointed Me (the Anointed One, the Messiah) to preach the good news (the Gospel) to the poor; He has sent Me to announce release to the captives and recovery of sight to the blind, to send forth as delivered those who are oppressed (who are down trodden, bruised, crushed, and broken down by calamity), To proclaim the accepted and acceptable year of the Lord (the day when salvation and the free favors of God profusely abound)."

(Luke 4:18, 19)

CHAPTER 1
KNOWING OUR HUMAN FRAILTY

You and I have been created in God's image. That means that we have the same nature as He does, which enables us to think, reason, make decisions, and express ourselves. God's nature is what gives us our creativity, our abilities, our intelligence, and our emotions. The Bible says that we are made from dust and return to dust when we pass away **(Ecclesiastes 3:20).**

This is a very strong way of telling us that we are most certainly created beings who cannot sustain ourselves. God literally breathed His life into Adam, the first human created **(Genesis 2:7).**

"For it was in Him (Christ) that all things were created, in heaven and on earth, things seen and things unseen, whether thrones, dominions, rulers, or authorities; all things were created and exist through Him (by His service, intervention) and in Him and for Him." **(Colossians 1:17)**

God created us in love because He is love and all that is pure and holy. We only know how to love because He first loved us; His nature is within us. Our personality, our likes and dislikes, our quirks: all have been given to us uniquely by a very personal and devoted Creator. Our strengths, our weaknesses, our successes, and our failures all make us the individuals that we are. But here's the rub! We cannot successfully manage ourselves, nor can we do anything without Him because that is the way He made us. We are dependent upon Him for our breath, our heartbeat, and everything else.

"And He Himself existed before all things, and in Him all things consist
(cohere, are held together)."

(Colossians 1:18)

"John answered, A man can receive nothing (he can claim nothing, he can take unto himself nothing) except as it has been granted to him from heaven."

(John 3:27)

"I am the Vine; you are the branches. Whoever lives in Me and I in him bears much fruit. However, apart from Me (cut off from vital union with Me), you can do nothing."

(John 15: 5)

So, we know that God made us in His image, and that image is loving, capable, intelligent, and strong. We also know that we as humans are very often not any of these things! We struggle to love ourselves and others; we feel very weak and incapable at times, often wondering how we are going to make it. The older I get, the more aware I am of my frailty and weakness physically. I have times that some would call "senior moments" that really make me shake my head at myself. Many times, we can laugh it off, but at other, more sober times, I think we ask ourselves, "Am I enough?" We wonder how to keep "it" all together, to manage to provide for ourselves and our families emotionally, keep our health strong, to continue to have enough financially. It is all we can do to feel confident when looking to the future.

For myself, I have learned (and *am still learning)* to lean on Jesus for what I need. I want to encourage you to do the same. In order to make this critical transition, I think it is necessary to identify our weaknesses our failings and to really accept the truth of who we are!

Admit it, right now, you're probably leaning towards the thought that you are fine; you aren't any "worse" than anybody else. Maybe you've taken care of yourself all your life, and you have a pretty good life. But, do you have a really deep, satisfying joy inside? Are your relationships healthy, or are they struggling? We all have things with which we struggle. Do you have anxiety and stress that keep you awake at night or possibly weigh so heavily that you struggle to stay "happy," and it affects your mood and the way you treat others? Are you burdened by unforgiveness towards others? Do you wish that what other people do didn't bug you so much? Believe me, I get it! If we dig far enough down, we will come face to face with a lot of junk.

(Jesus said): "The thief comes only in order to steal and kill and destroy. I came that they may have and enjoy life, and have it in abundance (to the full, till it overflows)."

(John 10:10)

"Come to Me, all you who labor and are heavy-laden and overburdened, and I will cause you to rest. (I will ease and relieve and refresh your souls). Take My yoke upon you and learn of Me, for I am gentle (meek) and humble (lowly) in heart, and you will find rest (relief and ease and refreshment and recreation and blessed quiet) for your souls."

(Matthew 11:28,29)

Did you know that Jesus is so easy to hang out with? Did you know that Jesus is crazy about you, concerned about you, determined to help you? He also told people that if they had seen Him, they had seen the Father, that He and the Father are One. So that means that God is also crazy about us, crazy enough to have sent Jesus here to die for us.

It is important to note that Jesus didn't come just to "patch us up" on the outside so we can cope in life still broken; He came *to seek and to save that which was lost* (Luke 19:10). Proverbs 23:7 tells us: "For as he (a man) thinks in his heart, so is he." And more strongly in Proverbs 28:26: "He who leans on, trusts in, and is confident of his own mind and heart is a (self-confident) fool, but he who walks in skillful and godly Wisdom shall be delivered." What we have is a heart problem! Each and every individual has the same condition unless and until they let Jesus in.

"The upright (honorable, intrinsically good) man out of the good treasure (stored) in his heart produces what is upright (honorable and intrinsically good), and the evil man out of the evil storehouse brings forth that which is depraved (wicked and intrinsically evil); for out of the abundance (overflow) of the heart his mouth speaks."

(Luke 6:45)

This, my friends, is the crux of God's *beckoning call to us*! He shows us in His Word that your heart, my heart, is truly *His desire.* He who designed us in such perfection and for a close, intimate relationship with us is ever-pursuing us. He could have scrapped the whole plan when Adam and Eve sinned and ushered destruction into the world. It would have seemed to be the logical result of such disobedience. But, the holy God of the universe was passionate about us, is passionate about us! No matter what, our behavior does not change how He views us. He *relentlessly pursues us out of His abundance of love, kindness, and mercy.*

As a segue into the next chapter, I am going to tell you a bit of my story: my brokenness, my journey, and my experiences. And how God beckoned to me with His love!

I grew up in the early '60s and '70s; it was a simpler time than today, with no computers, social media, or cell phones. But it was also a rough time for me to grow up; my personal experiences were oftentimes at odds with the lure of the "new age" that was dawning: women's rights and thumbing one's nose at authority. I was very young when America started to turn from the relative innocence of mom and pop raising their families to a more (so-called) enlightened way of life: taking authority over one's sexuality and the right to independence from societal norms. Teenaged rebellion became very much more normal. I got early and subtle messages about relationships (and myself) that I wouldn't process until much later in life. But they certainly shaped who I became, how I thought, and how I acted. Perhaps your story is similar.

My father left when I was just fourteen months old, and I never saw him again. This resulted in a real void in my life. I had two older sisters who were not close to me in age and a much younger brother. My mother did her very best, and looking back, I know that I was always loved and cared for deeply; her many sacrifices only became apparent to me as I grew and matured and understood what raising a family takes! But, the absence of my father and a mother who struggled in her own ways (mentally and financially) did cause me to have a lot of loneliness. Please note that none of what I share about my personal history is meant as fodder for someone to feel sorry for me! I have mourned that little girl, and Jesus has restored me and set me free from all the pain from my past! As a result of responding to His *beckoning call*, I have found freedom and an ability to share with others what used to be too painful to even 'look' at inside myself.

There was a family friend who molested me for a span of years when I was young, and that led to more isolation due to untold secrets. I never felt like I fit in well with other kids; much of this was because of the pain inside me that I could not voice to anyone else. Like many other kids, I was bullied and called derogatory names in school. I developed earlier than other girls and was an 'ingrown introvert'; this caused others to misunderstand me. As is common in humanity, we mock what we do not understand!

All of this "groundwork" led to an aloof, rude, dismissive young woman. I was sure (convinced) that others would dislike me, so I 'went first,' so to speak. I was obviously carrying a lot of pain in my soul! Once I was finished with school, I tried to move on; college turned out not to be for me. I left after just 6 months. I just never felt "special" or more than "ordinary," and that hurt!

I did come to believe in Jesus at a fairly young age; I grew up in a denominational church, learned doctrine and creeds, and was confirmed in the faith. Most of my experience and knowledge of God was gained educationally rather than spiritually. While I was convinced of the facts, I still lacked a true, intimate relationship with God. (for a fuller perspective on all of this, read my first book, The Weight of Expectations)

Although I have gone to church my whole adult life, it is just in the past fifteen years or so that I have really gotten to know *more about God. More about His love for me. More about my identity in Christ.* I believe the turning point for me was when I sensed the Holy Spirit stirring inside of me. The Holy Spirit lives inside of every believer, and He is our Helper.

"So too the (Holy) Spirit comes to our aid and bears us up in our weakness; for we do not know what prayer to offer nor how to offer it worthily as we ought, but the Spirit Himself goes to meet our supplication and pleads in our behalf with unspeakable yearnings and groanings too deep for utterance."

(Romans 8:26)

I am the apple of His eye; He is the Lover of my soul (this is true for every single believer!). The way I finally "got it" was by learning to *spend time in His Word (the Bible); spend time in His presence, spend time and just slow down.* Life choices and the end results are really about our priorities! Draw close to God, and He will draw close to you! (James 4:8) You know that you cannot have a relationship with anyone with whom you don't spend time.

CHAPTER 2
UNDERSTANDING GOD'S HEART

In order to fully grasp the depth of the Father's heart towards you, you have to be able to step back and really look at the beginning of everything. God created mankind as a companion that He could love and be good to and with whom He could enjoy fellowship. As I've mentioned, He created us in His image so we would be like-minded and have satisfying friendships between us.

"When I view and consider Your heavens, the work of Your fingers, the moon and the stars, which You have ordained and established, What is man that You are mindful of him, and the son of (earthbound) man that You care for him? Yet you have made him but a little lower than God (or heavenly beings), and You have crowned him with glory and honor."

(Psalm 8:3-5)

You are God's masterpiece! You were created *on purpose and for a purpose*. You might be thinking, "Wow! So, what happened there? I sure don't feel that special to God. I have no idea what my purpose is supposed to be!" Well, I can tell you that the truth is way simpler than we think it is! God may be complex and mysterious, but His plan is very straightforward. It is our humanness that gets in our way. It is the fallenness of humanity that has beclouded our understanding. But God has graciously provided the wisdom and knowledge that we need if only we will partake of it.

"Then they said, What are we to do, that we may be (habitually) working the works of God? (What are we to do to carry out what God requires?) Jesus replied, This is the work (service) that God asks of you: that you believe in the One Whom He has sent (that you cleave to, trust, rely on, and have faith in His Messenger)."

(John 6:28,29)

"But the just shall live by faith (My righteous servant shall live by his conviction respecting man's relationship to God and divine things, and holy fervor born of faith and conjoined with it); and if he draws back and shrinks in fear, My soul has no delight or pleasure in him."

(Hebrews 10:38)

If you aren't aware of the fallenness of mankind, then you won't fully understand the other side of the coin, God's redemptive purpose and plan. This is the crux of *his beckoning call.*

"Then while it is (still) called Today, if you would hear His voice and when you hear it, do not harden your hearts as in the rebellion."

(Hebrews 3:15)

We have an enemy, Satan, who is a fallen angel; he rebelled against God and His authority, so he was kicked out of heaven. Satan hates God and His creation, and he is the ruler of this present age. He does his level best to 'steal, kill, and destroy' us. (John 10:10). He is the power and instigator behind everything evil in the world!

"For we are not wrestling with flesh and blood (contending only with physical opponents), but against the despotisms, against the powers, against (the master spirits who are) the world rulers of this present darkness, against the spirit forces of wickedness in the heavenly (supernatural) sphere."

(Ephesians 6:12)

So then, if our enemy is supernatural, we need a stronger supernatural force (God) to help us fight our battles so that we are not overcome with fear and anxiety! Satan is the father of lies; he tells us all the wrong things. He is the root source of our fear, anxiety, dread, and discouragement.

"For God did not give us a spirit of timidity (of cowardice, of craven and cringing and fawning fear), but (He has given us a spirit) of power and of love and of calm and well-balanced mind and discipline and self-control."

(2Timothy 1:7)

So, if God has a great plan for us and an individual purpose for each of us, then what is it? How does this work? We tend to live our lives with a "me first" attitude because if we don't look out for ourselves, then who will? We decide how to live, what activities to participate in, who our friends are, and what our occupations will be. And, of course, God created us to think for ourselves and to operate in the world. What He did not intend, however, is that we would be independent of Him. We were created *in love* and for communion with our Creator. He is the life-giver; He is the One on Whom we are to rely.

> *"If any of you is deficient in wisdom, let him ask of the giving God (Who gives) to everyone liberally and ungrudgingly, without reproaching or faultfinding, and it will be given him."*

(James 1:5)

> *"The reverent fear and worship of the Lord is the beginning of Wisdom and skill (the preceding and the first essential, the prerequisite and the alphabet); a good understanding, wisdom, and meaning have all those who do (the will of the Lord). Their praise of Him endures forever."*

(Psalm 111:10)

> *"Behold, You desire truth in the inner being; make me therefore to know wisdom in my inmost heart."*

(Psalm 51:6)

I want to explore a concept of which I recently became aware. As I am writing this, it is the dawn of a new year, so it was recently Christmastime. Listening to Christmas music one day, I heard the reference to Santa as "naughty and nice"; how Santa doesn't come to bad little girls and boys. Isn't this exactly how the world operates? If you're not a "good" person, then don't expect anything from God! He is perfect and holy and certainly wouldn't have anything to do with a sinner. Many people use this reasoning to exempt themselves from hoping that there would be anything better for them because of *their behavior.* They have the idea that they can't approach God until they've straightened up or cleaned up their act. As I said earlier in this chapter, the truth of how God thinks and feels about us is so much simpler than we think it is; He is much kinder, more merciful, and more long-suffering than we know! The very plain truth is that we cannot clean ourselves up! If we humans had the ability to be perfect on our own, then there was no reason for Jesus to have gone through what He did on the cross.

"Inasmuch then as we have a great High Priest Who has (already) ascended and passed through the heavens, Jesus the Son of God, let us hold fast our confession (of faith in Him). For we do not have a High Priest Who is unable to understand and sympathize and have a shared feeling with our weaknesses and infirmities and liability to the assaults of temptation, but One Who has been tempted in every respect as we are, yet without sinning."

(Hebrews 4:14,15)

This tells us that Jesus also went through some "stuff" while He was here on earth. Although He withstood the temptations brought on by His trials, He understands how difficult it is. Jesus came as fully human and fully God. He knows that we need His help! He understands what it is like to be human. Isn't that comforting and encouraging?

Did you know that God is *for us*? That God wants to *help us*? Did you know that Jesus sits at God's right hand in heaven and that he is advocating for us, interceding on our behalf?

"And therefore, the Lord (earnestly) waits (expecting, looking, and longing) to be gracious to you; and therefore, He lifts Himself up, that He may have mercy on you and show loving-kindness to you."

(Isaiah 30:18)

"Therefore, He is able also to save to the uttermost (completely, perfectly, finally, and for all time and eternity) those who come to God through Him, since He is always living to make petition to God and intercede with Him and intervene for them."

(Hebrews 7:25)

I hope that by now, you are convinced that God's heart toward you is kind and his intentions for you are and have always been good! I also hope that as we continue in this book, God's love for you will be *felt and known* by you and that you will hear *His beckoning call.*

"For I know the thoughts and plans that I have for you, says the Lord, thoughts, and plans for welfare and peace, and not for evil, to give you hope in your final outcome."

(Jeremiah 29:11)

CHAPTER 3

RECOGNIZING GOD'S PRESENCE

<u>PRESENCE:</u> (A) The Fact Or Condition Of Being Present; Existence, Occurrence; Immediate Surroundings, Or Vicinity Within Close View.

It is one thing to recognize the presence of an object that has a physical form: there is a chair in the room, there is a mountain in the distance, or there is a house a mile away. We may not immediately see the last item I listed, but we know it exists because we've seen it in the past, or we can assume its existence because we know there are houses within the city limits. So, it is safe to say we can know that unseen objects have a presence, even if we are not in their vicinity.

Likewise, God had a physical presence. Jesus came to the earth and manifested as a human being. He had a physical presence here among us. Obviously, anyone who interacted with Jesus knew that he existed because they could see him. And, we who believe Christ came to earth know that he did live here once, although he has died like every other human who lived that long ago.

"And the Word (Christ) became flesh (human, incarnate) and tabernacled among us; and we (actually) saw His glory, such glory as an only begotten son receives from his father, full of grace and truth."

(John 1:14)

There is another level of knowing some*thing's* presence, the intangible. If you have a healthy relationship with someone, you know their love is present within that relationship. Their love for you is encompassed within your connection as friends, kin, etc. The special bond you have with that person is unique and real, yet unseen.

PRESENCE: (B) An Influence Or A Supernatural Or Divine Spirit Felt To Be Present

The dictionary is telling us here that the definition of presence also applies to a supernatural or divine spirit. That is harder to grasp because the spirit realm is truly unseen. Many of us have had little to no experience spiritually. Even that love I talked about between two people has physical form and expression to it to some degree. So, is it a stretch for you to understand God's presence here among us on earth? This is a critical concept if we are to have any relationship with the living God while we are here on earth.

"And we are setting these truths forth in words not taught by human wisdom but taught by the (Holy) Spirit, combining and interpreting spiritual truths with spiritual language (to those who possess the Holy Spirit). But the natural, nonspiritual man does not accept or welcome or admit into his heart the gifts and the teachings and the revelations of the Spirit of God..."

(1 Corinthians 2:13,14)

"However, I am telling you nothing but the Truth when I say it is profitable (good, expedient, advantageous) for you that I go away. Because if... I go away; I will send Him (Holy Spirit) to you (to be in close fellowship with you)....when He, the Spirit of Truth, comes, He will guide you into all the Truth."

(John 16:7,13)

Before we continue, I want to say be cautious when responding to and interpreting the spirit realm. Remember that Satan is also a spirit, and God cautions us that Satan will masquerade as a spirit of the light (one of God's own heavenly beings). (2 Corinthians 11:14) When Satan was kicked out of heaven, he took with him a third of the angels who were also rebellious and followed Satan. We must have a clear-headed understanding of these forces and the humility to know that we must be cautious about what we glean. Any question about what you are hearing or sensing should be presented to God; He is faithful in guiding us rightly. A trusted Christian friend or leader is also a good source to discern truth.

<u>RECOGNIZE:</u> (A) To Know By Some Detail, As Of Appearance; Identify

We should know and recognize that God is good; all His ways are righteous. He is loving, kind, gracious, merciful, and faithful! Not only is He *only good*, but everything that God is, is constant. He never changes, not in His thoughts and intentions toward us, nor in His plans and purposes. We can trust God; He is Who He says He is. He is worth getting to know because He never disappoints.

We could say that we recognize God's presence because we know that He is here in our midst. God is everywhere all the time, and in fact, we cannot hide from or escape His presence. We could also say that we recognize His presence by seeing the good that is in the world: sunrises, newborn babies, grand and breathtaking scenery, old couples who've been together for decades, random acts of kindness. We can 'identify' things that point out the presence of a good God. But simply knowing about God on this level doesn't truly help us or anybody else.

<u>RECOGNIZE:</u> (B) To Be Aware Of The Significance Of; To Acknowledge The Existence, Validity, Authority, Or Genuineness Of; To Acknowledge As Having Rights

God is, and has been, *beckoning to us* all our lives; He is aware of and involved in life here on earth! He wants to raise our consciousness of His presence but also to encourage and invite us to *come closer.* He is loving and wants to love us into wholeness. He wants to help us by giving us wisdom, direction, and clarity. He has been a witness to everything wrong that has hurt you; He knows where the broken pieces are, and He alone has the ability to heal and restore you!

And, to address your concerns, yes. God knows all the wrongs you have done, also. He knows every word in our mouths that is *yet unspoken!* And He loves us anyway; He longs to help us. He knows why you do what you do, the messed-up stuff that you just cannot seem to shake.

He knows that you act out of pain and brokenness. His *beckoning call* is saying, "That's okay, I'm here to help!"

"For God did not send the Son into the world in order to judge (to reject, to condemn, to pass sentence on) the world, but that the world might find salvation and be made safe and sound through Him."

(John 3:17)

"The thief comes only in order to steal and kill and destroy (Jesus said). I came that they may have and enjoy life, and have it in abundance (to the full, till it overflows)."

(John 10:10)

If we want to know God, and know the Truth, then we must read His Word, the Bible. Everything God says is Truth, meaningful, and relevant to us. Living an overcoming life full of hope, peace, and joy begins with knowing God. The Bible says that the knowledge of God is the beginning of wisdom.

"The mind of him who has understanding seeks knowledge and inquires after and craves it..."

(Proverbs 15:14)

"Every word of God is tried and purified; He is a shield to those who trust and take refuge in Him."

(Proverbs 30:5)

Another critical element of growing in our knowledge of God is fellowship with other believers. Jesus commanded us to gather and be with one another, to love one another as we do ourselves. There is strength in numbers; we need each other's encouragement, understanding, and sharing of the ups and downs in our lives. Many people have been turned off "church" because of hurtful experiences. I have had a few negative experiences myself! What we cannot, must not, do is let other people's weaknesses and failures in loving and accepting us drive us away from the strength and support of being with other believers. If you have been wounded in this way, please prayerfully consider a way forward *into God's grace and out of bitterness.*

"And let us consider and give attentive, continuous care to watching over one another, studying how we may stir up (stimulate and incite) to love and helpful deeds and noble activities, Not forsaking or neglecting to assemble together (as believers), as is the habit of some people...."

(Hebrews 10:24,25)

In closing, I'd like to encourage you to just *receive* what God has to offer. He offers it freely and universally. There is no one who is exempt from His grace, mercy, and forgiveness.

"In all your ways know, recognize, and acknowledge Him, and He will direct and make straight and plain your paths. Be not wise in your own eyes; reverently fear and worship the Lord and turn (entirely) away from evil. It shall be health to your nerves and sinews and marrow and moistening to your bones."

(Proverbs 3:6-8)

CHAPTER 4
HOW TO HEAR GOD'S VOICE

1. RELATIONSHIP: Connection Or Kinship

We know that all relationships involve communication; healthy, life-giving relationships require clear, mutual communication between parties. This should include the freedom to express oneself with confidence that the hearer will respect what we say and respond in kind. Unfortunately, we have all experienced or witnessed dysfunctional relationships due to a lack of respect or openness from one or both parties.

"To the one only God, our Savior through Jesus Christ our Lord, be glory (splendor), majesty, might and dominion, and power and authority, before all time and now and forever (unto all the ages of eternity). Amen."

(Jude 25)

How and whether we respond to God's authority will determine whether we have an actual relationship with God. Because sin is inherent in the world, we are all separated from God unless and until we are brought to repentance and, thus, to salvation through Jesus Christ. This includes every person, bar none. Jesus (and only Jesus) *is the Way, the Truth, and the Life.*

"God understands the way (to Wisdom), and He knows the place of it (Wisdom is with God alone)... to man He said, Behold, the reverential and worshipful fear of the Lord-that is Wisdom; and to depart from evil is understanding."

(Job 28: 23,28)

FEAR: Respectful Dread; Awe; Reverence; To Feel Reverence Or Awe For

We must understand that while God is capable of wrath when responding to evil, He is a loving God. The 'fear' of which we speak is awe and respect for His authority and power, an understanding that He is to be obeyed and what He says is to be taken seriously. His heart towards you is for rescue, redemption, and ultimate restoration. He longs to have a healthy relationship with you! He wants to call us His friend. At the same time, He is a holy God; our sins must be dealt with before He can commune with us. Jesus did that for us on the cross. Thank You, Jesus! (see John 3:16,17)

HEAR: To Perceive Or Sense; To Take Notice Of; Pay Attention To

Although God does 'speak' to His children here on earth, the main way He communicates is through His written word (the Bible). When I say speak, I don't mean audibly (although He can do it that way if He chooses to); God's Holy Spirit lives inside of each believer, and the Spirit absolutely does talk to us and commune with us.

"But the Comforter (Counselor, Helper, Intercessor, Advocate, Strengthener, Standby), the Holy Spirit, Whom the Father will send in My name (in My place, to represent Me and act on My behalf), He will teach you all things. And He will cause you to recall everything I have told you."

(John 14:26)

So, the first step in being able to hear from God is to come into a relationship with Him through Jesus Christ. Once we do that, the Holy Spirit lives inside of us forever. You read in the last Scripture that I shared above that the Spirit is our *advocate, intercessor, and helper*, among other wonderful personalities.

"Jesus answered him, I assure you, most solemnly I tell you, that unless a person is born again (anew, from above), he cannot ever see (know, be acquainted with, and experience) the kingdom of God."

(John 3:3)

"But you are not living the life of the flesh, you are living the life of the Spirit, if the (Holy) Spirit of God (really) dwells within you (directs and controls you). But if anyone does not possess the (Holy) Spirit of Christ, he is none of His (he does not belong to Christ, is not truly a child of God)."

(Romans 8:9)

2. **OBEDIENCE: Willingness To Submit To The Instructions Or Orders Of One In Authority Or Control**

"A new heart will I give you, and a new spirit will I put within you, and I will take away the stony heart out of your flesh and give you a heart of flesh. And I will put My Spirit within you and cause you to walk in My statutes, and you shall heed My ordinances and do them."

(Ezekiel 36:26-27)

The second and equally critical step to hearing God's voice is *having a heart set toward obedience.* Notice I did not mention perfection; I did not mention performance. God deals with us through our heart, our inner man. Having a willingness to obey is critical to pressing in knowing God and *hearing His voice!* If God has the heart, He can definitely change the behavior over time (with our cooperation).

Although our outward actions are important, nobody changes effectually by a force of will-theirs or anybody else's! Everything we do is motivated by our thoughts and what we believe. We must realize that *how we think, what we believe, and our heart attitude(s)* all will need to change.

It is important to know that God knows you are not, and will not be, perfect while here on earth. What He longs for and expects is an honest, open, sincere relationship with you. This takes perseverance and faithfulness. When you stumble or fail (and you will), *don't throw in the towel!* I'll say it again: what God thinks of us and expects of us is so much simpler than we think it is!

> *"Do not be conformed to this world (this age), (fashioned after and adapted to its external, superficial customs), but be transformed (changed) by the (entire) renewal of your mind (by its new ideals and its new attitude), so that you may prove (for yourselves) what is the good and acceptable and perfect will of God, even the thing which is good and acceptable and perfect (in His sight for you)."*

(Romans 12:2)

Studying God's Word is the single most effective way to know God. If you avail yourself of God's Truth, you will understand God's ways, and they are so much higher and better than our ways. The important attitude with which to approach God's Word is receptivity and the lack of fear of condemnation. Know and understand that God *wants you to know Him.* The Bible is for His children to know and understand their Father!

> *"Therefore, (there is) now no condemnation (no adjudging guilty of wrong) for those who are in Christ Jesus, who live (and) walk not after the dictates of the flesh, but after the dictates of the Spirit."*

(Romans 8:1)

Of course, we are going to run into truths that make us squirm; we are going to see how far short we fall *on our own,* and that is the whole point of the Truth. Not to press us down and make us feel like worms but to show us what is possible for us if we surrender to God. Remember this: God created you for communion with Him; Jesus came and redeemed you; the Holy Spirit lives *inside of you to help you!* Where in all of this would lie a trap? How, believing this, could we deny that God loves us? How could we not believe that God is for us?

"And therefore, the Lord (earnestly) waits (expecting, looking, and longing) to be gracious to you; and therefore, He lifts Himself up, that He may have mercy on you and show loving-kindness to you, for the Lord is a God of justice. Blessed (happy, fortunate, to be envied) are all those who (earnestly) wait for Him, who expect and look and long for Him (for His victory, His favor; His love, His peace, His joy, and His matchless, unbroken companionship)!"

(Isaiah 30:18)

3. PRAYER: Talking To Your Beloved

"Let us then fearlessly and confidently and boldly draw near to the throne of grace (the throne of God's unmerited favor to us sinners), that we may receive mercy (for our failures) and find grace to help in good time for every need (appropriate help and well-timed help, coming just when we need it)."

(Hebrews 4:16)

"Rejoice in the Lord always (delight, gladden yourselves in Him)... Do not fret or have any anxiety about anything, but in every circumstance and everything, by prayer and petition (definite requests), with thanksgiving, continue to make your wants known to God. And God's peace (shall be yours...)"

(Philippians 4:4,6,7)

I hope that you are beginning to see that *your heart truly is God's desire!* Scripture makes it so unmistakable. I just adore God's word! It truly has the power to save our souls (James 1:21). Studying what God says about me encourages me to want to talk to Him. There is nothing hard, harsh, sharp, or pressing about Jesus. Let me encourage you to reach out in prayer to God. He understands our jumbled words; He understands our silences. He hears the faintest cries of our hearts. He rejoices over you with singing!

"The Lord your God is in the midst of you, a Mighty One, a Savior! He will rejoice over you with joy; He will rest (in silent satisfaction), and in His love, He will be silent and make no mention (of past sins or even recall them); He will exult over you with singing."

(Zephaniah 3:17)

"Only aim at and strive for and seek His kingdom, and all these things shall be supplied to you also. Do not be seized with alarm and struck with fear, little flock, for it is your Father's good pleasure to give you the kingdom!"

(Luke 12:31,32)

I'm using the Scriptures to show you and to reassure you that God loves you, wants you, and longs to hear from you. If you're not sure "how" to talk to God, who seems so far away, don't feel awkward. Even Jesus's disciples asked Him to teach them how to pray! I'm going to share that passage (Luke 11: 1-4) with you here. The bulleted sentences are merely for clarification.

“Then He was praying in a particular place; and when He stopped, one of His disciples said to Him, Lord, teach us to pray, (just) as John taught his disciples. And He said to them, when you pray, say:

- *Our Father Who is in heaven, hallowed (holy, sacred) be Your name*
- *Your kingdom come, Your will be done (held holy and revered) on earth as it is in heaven.*
- *Give us our bread daily (provision for today)*
- *And forgive us our sins, for we ourselves also forgive everyone who is indebted to us (who has offended us or done us wrong)*
- *And bring us not into temptation but rescue us from evil.*

Here, Jesus is showing them what God cares about: that we acknowledge Him, that we desire to do His will and to see His kingdom (His holiness) reflected in this world, that we come to Him for our daily needs, that we seek forgiveness and understand that forgiveness from us is important; and, that we ask Him for protection from the evil one. God wants us to depend on Him! He wants us to know and to acknowledge that we need Him. 24/7, we need Him.

“Lean on, trust in, and be confident in the Lord with all your heart and mind, and do not rely on your own insight or understanding. In all your ways, know, recognize, and acknowledge Him, and He will direct and make straight and plain your paths. Be not wise in your own eyes; reverently fear and worship the Lord and turn (entirely) away from evil. It shall be health to your nerves and sinews and marrow and moistening to your bones.”

(Proverbs 3:5-8)

CHAPTER 5
THE HEART'S RESPONSE

"He has made everything beautiful in its time. He has also planted eternity in men's hearts and minds (a divinely implanted sense of a purpose working through the ages which nothing under the sun but God alone can satisfy)..."

(Ecclesiastes 3:11)

"That is why I (Jesus) told you that you will die in (under the curse of) your sins; for if you do not believe that I am He (Whom I claim to be if you do not adhere to, trust in, and rely on Me), you will die in your sins."

(John 8:24)

God loves you so very much! I am sure that by now, you have had the opportunity to sense His beckoning call to *your* heart. I assure you that God does *delight in you, and He longs to love you.* I pray that whatever doubts you may still harbor, and whatever the reasons are that you still struggle with your decision, all of that may dissolve when you take the time to ponder in your heart what I've said. May God's Truth soak down into your inner man and set you free!

"As He said these things, many believed in Him (trusted, relied on, and adhered to Him). So, Jesus said to those Jews who had believed in Him, If you abide in My word (hold fast to My teachings and live in accordance with them), you are truly My disciples. And you will know the Truth, and the Truth will set you free."

(John 8: 30-32)

One of the main things that helped me progress from knowing *about* God to wanting to *know God* was disciplining myself to spend time with Him regularly and consistently. We must take the inkling of faith we have and plant it in the promises of God. We must be willing to "check it out and find out" if we are going to make any progress. God has already done the 'work' of salvation for us; He has given us His inspired Word (the Bible); He has given us other believers to share their testimonies. But ultimately, the "success," if you will, of our God journey is up to us. The following Scripture verses are very dear to me, and I believe they illustrate the way to a joyful, successful, peace-filled life!

"Trust (lean on, rely on, and be confident) in the Lord and do good; so shall you dwell in the land and feed surely on His faithfulness, and truly you shall be fed. Delight yourself also in the Lord, and He will give you the desires and secret petitions of your heart."

(Psalm 37: 3,4)

Remember that once we put our faith in Jesus, His Holy Spirit is given to us to dwell inside of us. The Holy Spirit reveals all Truth to us (John 14:17, 26) and guides us in our journey. But we must have patience and let the Spirit work; God is never in a hurry. Once you've planted your faith in God, give it time to grow. Know that *it is not about your performance!* It is a relationship with God the Father through His Son, Jesus Christ.

"But the fruit of the (Holy) Spirit (the work which His presence within accomplishes) is love, joy (gladness), peace, patience (an even temper, forbearance), kindness, goodness (benevolence), faithfulness, Gentleness (meekness, humility), self-control (self-restraint, continence). Against such things, there is no law."

(Galatians 5:22,23)

"A time will come, however, indeed it is already here when the true (genuine) worshipers will worship the Father in spirit and truth (reality); for the Father is seeking just such people as these as His worshipers."

(John 4:23)

How To Receive Jesus As Your Lord And Savior

The most important relationship of your life is a personal relationship with Jesus Christ. If you would like to receive Him as your Lord and Savior and enter into the greatest relationship you have ever known, please pray the prayer below.

Father,
You loved the world so much that You gave Your only begotten Son to die for our sins so that whoever believes in Him will not perish but have eternal life. Your Word says we are saved by grace through faith as a gift from You. There is nothing we can do to earn salvation.
I believe and confess with my mouth that Jesus Christ is Your Son, the Savior of the world. I believe He died on the cross for me and bore all of my sins, paying the price for them.
I believe in my heart that You raised Jesus from the dead and that He is alive today.
I am a sinner, and I am sorry for my sins. I ask You to forgive me. By faith, I receive Jesus Christ now as my Lord and Savior. I believe that I am saved and will spend eternity with You!
Thank You, Father.
I am so grateful! In Jesus' name,
Amen.

Made in the USA
Monee, IL
13 September 2026

63177764R00020